TROPICAL FISH

TROPICAL FISH

FROM STINGRAYS TO CLOWN FISH

DAVID ALDERTON

First published in 2025

Copyright © 2025 Amber Books Ltd

All rights reserved. No part of this publication may be reproduced, stored in a retrieval system, or transmitted in any form or by any means, electronic, mechanical, photocopying, recording, or otherwise, without prior written permission of the copyright holder.

Published by
Amber Books Ltd
United House
North Road
London N7 9DP
United Kingdom

www.amberbooks.co.uk
Facebook: amberbooks
YouTube: amberbooksltd
Instagram: amberbooksltd
X(Twitter): @amberbooks

ISBN: 978-1-83886-551-1

Project Editor: Anna Brownbridge
Designer: Keren Harragan
Picture Research: Adam Gnych

Printed in China

Contents

Introduction	6
Freshwater	8
Marine	96
Invertebrates	184
Picture Credits	224

Introduction

Fish represent one of the most diverse and fascinating groups of vertebrates, with their distribution extending over the entire planet, from freezing polar seas to the waters of the tropics, with a few even occasionally venturing on to land. Certain fish swim huge distances through the world's oceans, while others spend their entire lives in tiny pools. Their breeding habits are equally fascinating and varied, as fish can either give birth to live young or reproduce by means of eggs.

A surprising number actually look after their offspring through the early stages of life, even producing food for them.

Fish also display an incredible range in size, from the dwarf goby (*Paedocypris progenetica*) found on the Indonesian islands of Sumatra and Bintan, measuring just 7.9mm (0.31in) when mature, up to the gigantic yet harmless whale shark (*Rhincodon typus*) which is known to reach a length of at least 18.8m (61.7ft), and can weigh 21.5 tonnes!

ABOVE:
Creating confusion
In some cases, fish can be masters of disguise, as exemplified by the ornate ghost pipefish (*Solenostomus paradoxus*) seen here, drifting like seaweed in the water. This species occurs around the edges of reefs in the western Pacific Ocean, and also in the Indian Ocean.

OPPOSITE:
Extra help
Many fish occur in groups called shoals, as with these Asian glass catfish (*Kryptopterus* species), which have transparent bodies. They are normally found in fairly murky waters, and so have developed the hair-like barbels, which help them to navigate here and locate their prey.

Freshwater

Fish come in a huge variety of shapes and sizes. There are currently reckoned to be around 30,000 different species occurring in waters throughout the world. There is a fairly even split between those that are found in freshwater, and others that occur in marine environments. Very few fish have the ability to transition from freshwater surroundings to the sea, however, because their physiology has adapted to the chemistry of their surroundings. Those found in freshwater have large, very efficient kidneys that retain body salts in their bodies. The reverse applies in the case of fish that live in the sea, where they are swimming in a concentrated salt solution. This then leaves them at risk of becoming dehydrated, which is the reverse situation of those found in freshwater. As a result, they are forced to drink large amounts of saltwater, with their very efficient kidneys helping to minimize water loss from the body. When it comes to water temperature, some fish are well adapted to live at the extremes. The desert pupfish (*Cyprinodon macularius*), found in parts of the United States (where it is now an endangered species), can survive in water temperatures as low as 4°C right up to 45°C (39–113°F), and at high salt levels too. At the other extreme, the marine Antarctic blackfin icefish (*Chaenocephalus aceratus*) possesses a form of antifreeze in its blood, to prevent it from dying in these bitterly cold waters.

OPPOSITE:
Swimming power
Fish rely on their fins not just to help them swim, but also to maintain their balance in the water.

Shaped by their environment
The body shape of fish can be very variable, often being directly linked to their habitat. Angelfish (*Pterophyllum* species) have tall, narrow bodies that enable them to swim or retreat easily through reedy areas where they can also hide, with their striped patterning providing them with camouflage there.

Breathing mechanisms
Most fish depend on their gills, located on each side of the head, for breathing purposes, extracting oxygen from the water, but a few, such as African lungfish (*Protopterus* species), are also able to gulp down atmospheric air directly from the surface.

BOTH PHOTOGRAPHS:
Hunting techniques
Archerfish (*Toxotes* species) prowl just below the water surface, looking for small invertebrates on branches above. When a target is spotted within range, the archerfish fires a jet of water from its mouth, hopefully knocking the invertebrate into the water, where it can be easily snapped up.

Determined hunters
Arowanas or bony-tongues (*Osteoglossinae*) also hunt at the surface. These large fish, growing to about 0.9m (3ft) long, are capable of leaping up to 1.8m (6ft) out of the water, targeting unwary birds and other creatures within reach, and then grabbing them in their powerful jaws.

LEFT AND ABOVE TOP:
A unique feature
The primitive ornate bichir (*Polypterus ornatipinnis*) is the only known vertebrate to possess lungs, while lacking a trachea (airway) which would allow it to breathe through its mouth. Instead, it inhales and exhales through a pair of spiracles, similar to blowholes, on the top of the skull.

ABOVE BOTTOM:
Markings with purpose
Also found in Africa, *Astatotilapia burtoni* has a distinctive mating behaviour. A dominant male becomes more colourful and releases pheromones (chemical messengers) in urine to attract a mate. She will ultimately nibble at the egg-like spots on his anal fin when ready to spawn.

LEFT:
Charged up
Black ghost knifefish (*Apteronotus albifrons*) occur in the Amazon region. These fish have electrical organs and sensors, which together alert them to the presence of prey.

OPPOSITE TOP AND BOTTOM:
A key difference
A bristlenose catfish (*Ancistrus* species), so-called because of the bristles apparent on the forehead of the male, as seen above, compared with the female below.

ALL PHOTOGRAPHS:
Blind cave fish
The Mexican tetra (*Astyanax mexicanus*) represents a cave-dwelling population of fish which have lost the use of their eyes, relying on the sensory lateral line running down the midline on each side of the body to avoid danger and find food in their dark environment.

Spotting the differences
It can be possible in some cases, as with the Congo tetra (*Phenacogrammus interruptus*), to distinguish the sexes, not necessarily just on grounds of their colouration, but also because of differences in the shape of their fins. The male here in front is also more brightly coloured.

New variants
A number of freshwater fish are very popular as aquarium occupants, such as the discus (*Symphysodon discus*), and these particular fish have been selectively bred in a wide range of colours, which are far more striking than the original wild forms from which they were created.

Parental care
Unusually among fish, discus form a strong pair bond, and the adults produce a protein-rich secretion, called 'discus milk', on the sides of the body to nourish their young when they hatch. They also watch over their offspring in these early stages, carefully chaperoning them.

BELOW:
A clean start
Discus prove to be devoted parents, starting initially by cleaning the area where the female deposits up to 400 eggs. Rockwork or submerged wood is often favoured as a spawning site. The eggs seen here are developing well, aside from the two whitish eggs, which are infertile.

OPPOSITE
Escaping danger
The young discus stay close to their parents for the first two weeks of their lives. If danger threatens, they are rapidly chaperoned to a suitable hiding place, often a small depression in the river bed or among vegetation, until the risk of attack passes.

ABOVE AND OPPOSITE:
An artificial creation
The remarkable flowerhead cichlid does not occur naturally. It has been created by cross-breeding other cichlids in aquarium surroundings. Males are characterised by their very large nuchal hump. This hump tissue develops in response to hormonal stimulus and contains a large amount of fat.

RIGHT:
Flying through water
The African butterflyfish (*Pantodon buchholzi*) lives close to the water's surface. The broad pectoral fins at the side of its body resemble a butterfly's wings.

LEFT AND BELOW:
A natural jewel
Unlike many African cichlids, the jewel cichlid (*Hemichromis bimaculatus*) occurs in rivers rather than Africa's Great Lakes. Its beauty means it is a popular aquarium fish.

OPPOSITE TOP AND BOTTOM:
Double vision
Four-eyed fish (*Anableps* species) have remarkable eyesight, with their eyes being effectively split in half. They can see both above and below the water's surface at the same time.

RIGHT AND OVERLEAF:
Hiding in plain sight
Not all fish are colourful. The Asian glass catfish (*Kryptopterus* species) has a transparent body, allowing it to merge very effectively into the background of its surroundings.

An aquarium favourite
Probably the most widely kept tropical aquarium fish in the world, the guppy (*Poecilia reticulata*) has been bred in a myriad of colours and fin forms. It is named after the Reverend Guppy, who sent specimens to the Natural History Museum in London from the island of Trinidad.

Telling them apart
There is a clear distinction between the sexes in the case of guppies. Males, as seen here, are much more colourful and smaller than females, which give birth to live young. They are mature by about seven weeks of age, and may live for up to two years.

ALL PHOTOGRAPHS:
A variable appearance
In spite of its name, the lemon cichlid (*Neolamprologus leleupi*) from Lake Tanganyika in Africa can vary from a predominantly rich shade of yellow through to brown.

ALL PHOTOGRAPHS:
Not that affectionate!
Kissing gouramis (*Helostoma temminckii*) are known to lock their protruding mouthparts together. This is not a sign of affection, though, but rather a test of strength between rivals, and therefore represents a display of stylized aggression with no harmful consequences in terms of injuries.

OVERLEAF BOTH PHOTOGRAPHS:
Easier eating
These fish, which are found in parts of southeast Asia, actually have an unusual jaw structure. They possess an extra joint here that serves to help them to feed more easily on vegetation growing on underwater rocks, using their tiny teeth to strip off the algae.

A snake-like appearance
Kuhli loaches (*Pangio* species) originate from parts of southeast Asia, inhabiting muddy or sandy areas of water. They will dig in the substrate in search of food such as small aquatic invertebrates and live in groups in the wild, although they do not actively swim in shoals.

The key sign
The Indian glassfish (*Parambassis ranga*) is another species which has a transparent body, allowing its skeleton to be seen. This individual can be recognized as a male, thanks to the black marking apparent on its dorsal fin, on the top of its body. They grow to about 7.5cm (3in) long.

LEFT TOP:
A wide choice
Mollies (*Poecilia* species) are found mainly in Central America. They have been widely bred for aquariums, with today's domestic strains bearing little resemblance to wild relatives.

LEFT MIDDLE:
Variations on a theme
Although mollies are now often brightly coloured, there is also a very popular strain called the black molly. In the wild, these fish often live in brackish (slightly salty) water.

LEFT BOTTOM:
Colourful breeds
Cross-breeding between different mollies, and also other closely related fish such as platies (*Xiphophorus* species), helps to explain the range of colours that exist today.

OPPOSITE:
Swollen shape
This colourful balloon molly is a lyretailed variant, distinguishable by the shape of its tail or caudal fin, which is longer at the top and bottom.

LEFT AND OVERLEAF:
Biological control
Mosquito fish (*Gambusia affinis*) originate from North America, but have been introduced to tropical areas in the past, in the hope they would assist in controlling malaria by eating mosquito larvae.

ALL PHOTOGRAPHS:
Shoaling fish
The stunning neon tetra (*Paracheirodon innesi*) is a shoaling fish found in the Amazon region, where it was first discovered in 1936. It has since become a hugely popular aquarium species.

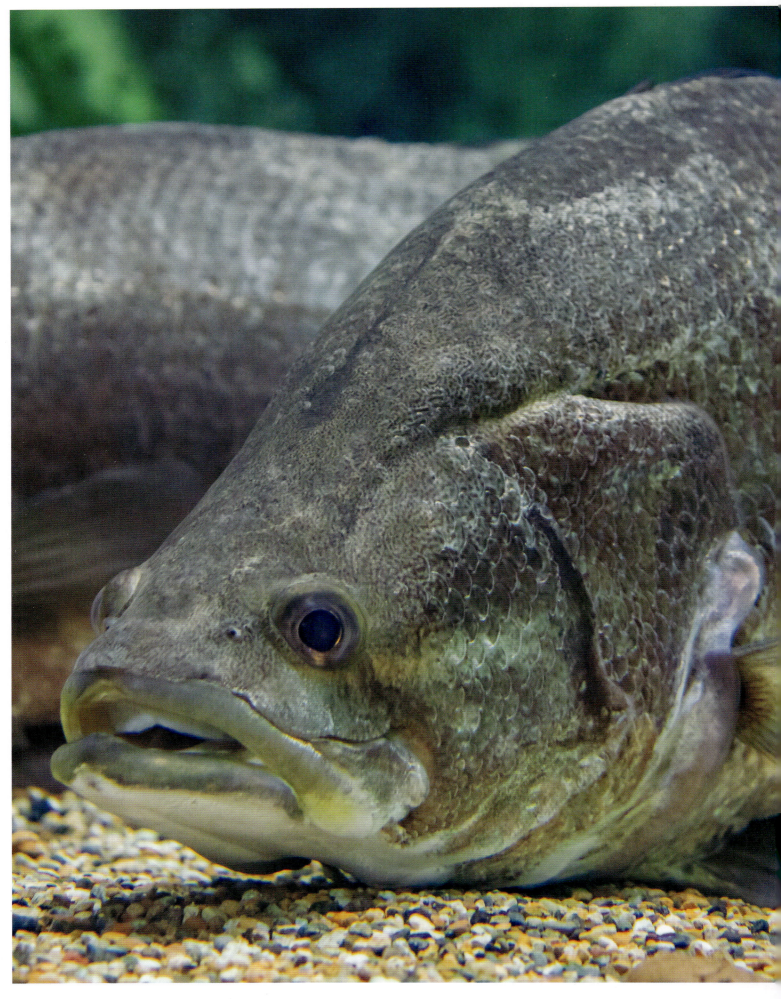

A deadly invader
The Nile perch (*Lates niloticus*) is a giant among freshwater fish, with the largest individuals growing up to almost 2m (6ft 7in) and weighing 200kg (440lb). It has been introduced to many African lakes as a food source, but can cause great environmental damage, preying on smaller fish.

ABOVE:
Seen from below
The underside of an ocellate river stingray (*Potamotrygon motoro*). These fish occur over a wide area of South America. Their disc-shaped body can measure 50cm (20in) wide and they may weigh 35kg (77lb).

LEFT:
A painful defence
The venomous barb of an ocellate river stingray seen in close-up, protruding from the upper surface of the tail. People can unfortunately tread on it when the ray is resting in shallow water.

OPPOSITE:
A different skeleton
Rays can be encountered both in freshwater and in the sea. They belong to the cartilaginous (rather than bony) group of fish. An ocellate river stingray is seen here.

Super-charged
The elephant-nose fish (*Gnathonemus petersii*) is so-called because of its trunk-like appendage. This, however, is not actually a nose, but part of the mouth, and helps these fish to use electric currents generated in the tail to track down the invertebrates that form their diet.

ALL PHOTOGRAPHS:
Perfect miniatures!
Ranking as one of the smallest freshwater fish, dwarf rasboras (*Boraras maculatus*) only grow to a maximum size of 2.5cm (1in) when adult. They originate from southeast Asia.

An infamous hunter
A red-bellied piranha (*Pygocentrus nattereri*), revealing its fearsome array of teeth: there are in fact over 30 different species of these notorious predatory New World fish. This particular species is found over a wider area than any other, and is primarily a scavenger, and also omnivorous.

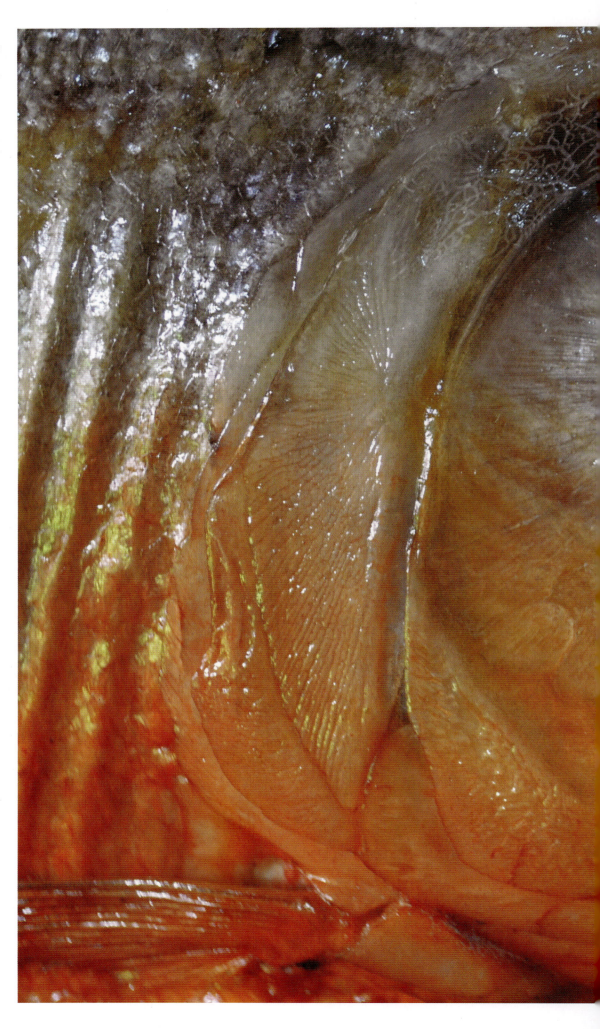

Strength in numbers
Piranhas associate in shoals which can number hundreds of individuals, and this can occasionally lead to fearsome feeding frenzies, with the water literally churning as the fish compete for food. More commonly, piranhas conceal themselves, ambushing other fish as they swim past them.

A deceptive appearance
The red-tailed black shark (*Epalzeorhynchos bicolor*) belongs to the cyprinid (carp) family, and is simply described as a 'shark' because of its body shape. It grows up to 20cm (8in) long. Originating from Thailand, it was feared to be extinct in the wild there until being rediscovered in 2011.

ABOVE BOTTOM:
Equipped to survive
The unusual reedfish or ropefish (*Erpetoichthys calabaricus*) lives in both fresh and brackish water in western and central parts of Africa. It has lungs as well as gills, so that it can breathe atmospheric air directly in localities where there is little oxygen in the water.

ABOVE TOP AND RIGHT:
A lively catfish
The mighty red-tailed catfish (*Phractocephalus hemioliopterus*) is one of the largest fish to be found in the Amazon region, growing up to 1.8m (6ft) long, and weighing 80kg (180lb). The long projections around the mouth, called barbels, help the catfish to hunt prey successfully in murky water.

BOTH ABOVE PHOTOGRAPHS:
A master of disguise
The South American leaf fish (*Monocirrhus polyacanthus*) closely resembles a leaf. The pointed tip on its lower jaw even resembles a leaf stalk! It reinforces this impression by drifting in the water at an angle, while being on the lookout for prey passing within easy reach.

RIGHT:
A popular choice
Rosy barb (*Pethia conchonius*) have a range extending from Afghanistan eastwards to Bangladesh in southern, subtropical parts of Asia. They are very popular aquarium species, thriving in small groups in these surroundings, and they can be persuaded to spawn without too much difficulty.

A powerful jumper
Found in northern parts of South America, the common hatchetfish (*Gasteropelecus sternicla*) lives near the water surface. It will use its pectoral fins to power itself out of the water to catch insects or escape danger. Its name comes from its body profile, which resembles that of a hatchet.

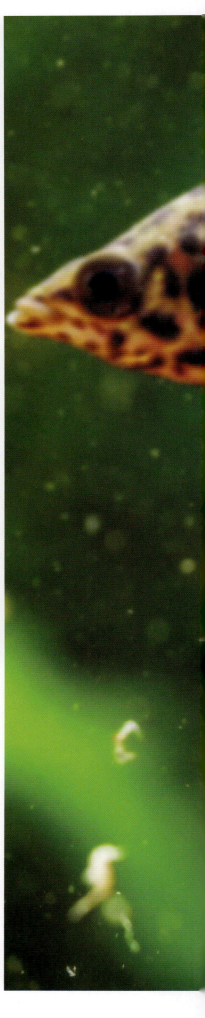

BOTH ABOVE PHOTOGRAPHS:
Duelling fish
Swordtails (*Xiphophorus hellerii*) are popular livebearing aquarium fish, now bred in many colours, which originated from Central America. They are so-called because of the sword-like projection evident on the bottom part of the caudal (tail) fin of males. Females lack this characteristic.

RIGHT:
On the move
The leopard bush fish (*Ctenopoma acutirostre*) is an African fish, found in the Congo River basin. It is a predatory species, feeding on smaller fish. Also known as the spotted climbing perch, it can walk short distances over land, breathing atmospheric air at this stage.

Changing locations
The spotted climbing perch has special auxiliary labyrinth organs located under the gill flap behind each eye. Being able to breathe in this way helps these fish to avoid becoming trapped in pools that may be drying up, and head instead to larger areas of water.

RIGHT AND OPPOSITE TOP:
A sedentary nature
Catfish as a group display a wide range of different body shapes. The twig catfish (*Farlowella* species) is a narrow-bodied, herbivorous South American species that relies on its stick-like disguise in the water.

OPPOSITE BOTTOM:
A different perspective
The African upside-down catfish (*Synodontis nigriventris*) spends most of its time swimming on its back. This helps it to catch insect life at the water surface more easily, and possibly confuse predators too.

87

Southeast Asian livebearers
Wrestling halfbeaks (*Dermogenys pusilla*) display a marked difference in the length of their jaws. Males also lock jaws when disagreeing, for up to 30 minutes, with the weaker individual ultimately breaking away. The larger females will give birth to up to 20 live young.

A deadly hunter
The zebra shovelnose (*Brachyplatystoma tigrinum*), also called the tiger-striped catfish, is a streamlined and highly predatory fish growing to about 90cm (36in) in length. It is a powerful swimmer at home in whitewater sections of river in northern South America, where it hunts other fish.

RIGHT AND OVERLEAF:
New forms
The long-finned zebrafish (*Danio rerio*) is a domesticated variant, while the use of this species in genetic research has led to the creation of transgenic zebrafish in red and other colours.

Marine

Some of the most colourful and beautiful fish in the world are to be found living on the world's coral reefs, along with some of the most bizarre and dangerous. There are just over 15,000 different species of marine fish known from the world's oceans, with more awaiting discovery, particularly in their depths. We are only just beginning to obtain insights into the fish that live there, in permanent darkness and at great pressure, thanks to advances in technology such as remote-control submersibles that have allowed us to penetrate to depths where divers could not survive. Although deep sea fish are not generally colourful, they come in strange forms, and some are capable of generating their own light in these murky depths.

This is not to find their way around, but rather to communicate with others of their kind or to attract prey, as in the case of anglerfish. They produce light thanks to luminescent bacteria within their bodies. Many deep sea fish undertake daily movements, heading up in the water column as darkness descends above, before returning to the depths before daylight dawns. Other fish roam the open oceans, being capable of travelling at speed here, to catch prey and escape predators themselves. Certain habitats, notably mangroves, provide cover and attract young fish early in life when they are especially vulnerable to predators, although once they grow older, they will then move back to coral reefs where they were spawned.

OPPOSITE:
Hiding away
The eye of a European angler or frog fish (*Lophius piscatorius*) seen in close-up. The relatively large size of the eyes improves the ability of the fish to see in darkened surroundings. The short appendages on their bodies resembling seaweed help to conceal these ambush hunters.

Lighting the way
Relatively little is known about deep sea marine fish like this anglerfish (*Diceratias pileatus*) which lives in the dark depths of the ocean down to a depth of at least 1430m (4692ft). Indeed, only recently has its range been extended from the Atlantic Ocean into Indo-Pacific waters too.

LEFT:
Defence in numbers
A shoal of sweetlips (*Plectorhinchus* species). By swimming closely together if threatened, the striped patterning of these fish works rather like that of zebras facing predators on the African plains. The stripes effectively merge together, making it harder for a predator to focus on an individual.

ABOVE:
A unique habitat
The Great Barrier Reef lies off the eastern coast of Australia, and is home to more than 1500 different species of fish. It consists of numerous smaller coral reefs, which extend collectively for over 2300 km (1400 miles) – further than any other reef system in the world.

Strength in numbers
Young barracudas (*Sphyraena* species) often form shoals. Note their streamlined body shape, which helps them to swim fast, while their silvery body colouration is hard for prey to detect. Barracudas will attack schools of fish, sometimes driving them into shallower water where they will be easier to catch.

ABOVE AND RIGHT:
A ferocious hunter
Barracudas usually become solitary as they grow older. They possess a fearsome array of teeth and hunt mainly by sight, growing to more than 1.65m (5.4ft) in length.

OPPOSITE TOP AND BOTTOM:
Leaving the water
Mudskippers (*Oxudercinae*) have an unusual body shape for a fish, and can spend extended periods out of the water, sometimes skipping or even jumping across the mud.

ABOVE AND RIGHT:
Equipped for the open ocean
Bigeye tuna (*Thunnus obesus*) may grow to over 2.5m (8ft) long, and weigh 180kg (397lb). They have excellent eyesight and can maintain their body temperature even in cold water.

OPPOSITE:
Equipped for the open ocean
Some of the most colourful marine fish inhabit the world's tropical reefs. This is a ribbon eel (*Rhinomuraena quaesita*), found in the Indo-Pacific region.

Hiding on the reef
The ribbon eel is one of the most slender-bodied fish of its type, and may have as many as 255 bones making up its vertebral column. This is a male; the larger females are typically yellow, and grow to a length of about 130cm (51in).

A stunning sight
This reef fish is described as a palette surgeonfish (*Paracanthurus hepatus*), because its black markings resemble those of an artist's palette, although it is also called the regal tang. The description of 'surgeonfish' comes from the sharp needle-like spine present at the base of the tail.

A body cleanse
Two bluestreak cleaner wrasses (*Labroides dimidiatus*) encircle a pufferfish, ready to remove any parasites from its body which will provide them with food. They congregate in special 'cleaner stations' on the reef, where fish of different types come to be treated, leaving the wrasses unmolested.

Changing sex
Bluestreak cleaner wrasses are unusual, because if a dominant male dies, and there is no other male to take over, then a mature female will develop into a male, assuming this role. These fish can be found on reefs from the Red Sea to the western Pacific.

A deadly killer
A bull shark (*Carcharhinus leucas*) swimming off the coast of Mexico. These sharks are highly unusual in that they can move into fresh water and may swim considerable distances in such surroundings, having been encountered over 1100km (1770 miles) up the Mississippi River from the sea.

Solitary hunters
Few sharks are more aggressive than bull sharks, and they are often found in shallow coastal areas, where they represent a danger to people. They are apex predators, hunting other fish, as well as other marine creatures, including dolphins and turtles, being opportunistic hunters.

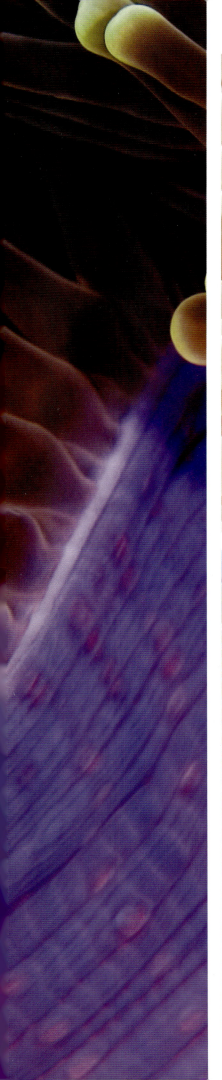

ALL PHOTOGRAPHS:
Living together
Anemonefish or clownfish (*Amphiprioninae*) possess a special mucus on their bodies, protecting them from the stinging tentacles of sea anemones, so they can retreat here if danger threatens. In return, they may bring food back for the sea anemone. This mutually beneficial arrangement is called symbiosis.

Unexpected protection
The clown triggerfish (*Balistoides conspicillum*) has protective spines along its back, behind the eyes. These can be raised if the fish is threatened, making it difficult for a predator to swallow triggerfish, although the spines are normally kept flat and largely hidden within a groove.

OPPOSITE TOP AND BOTTOM:
Increasing the odds
Fish frequently produce large numbers of eggs, because only a tiny proportion will survive through to adulthood. Different fish use different methods to influence the survival rates of their young; anemonefish deposit their eggs close to sea anemones, which will ultimately protect them.

ABOVE:
Parental care
Some fish lay fewer eggs but invest more effort in parental care through the critical early stages after spawning occurs. In the case of cardinalfish (*Apogonidae*), the eggs are collected at this stage by the male fish and carried in his mouth until the young hatch.

ABOVE:
Amazing appearance
The flame angelfish (*Centropyge loricula*) is stunningly coloured. It is encountered on tropical Pacific reefs, eating plankton when young, and then algae and crustaceans.

RIGHT AND FAR RIGHT:
Variations on a theme
Western beaked butterflyfish (*Chelmon marginalis*) (seen right) and a copperband butterflyfish (*C. rostratus*) have dark eyespots on the rear of the body to confuse would-be predators. Their long snouts allow them to probe around the reef for small invertebrate prey.

ALL PHOTOGRAPHS:
Staying in place
The royal gramma or fairy basslet (*Gramma loreto*) is seen on tropical reefs in the western Atlantic. When spawning, the male constructs a nest from algae for the eggs, which have tiny threads. These anchor onto the algae, preventing the eggs from drifting away.

BOTH PHOTOGRAPHS:
Taking to the air
Flying fish (*Exocoetidae*) live in tropical waters. They will leap out of the sea to escape pursuing predators like tuna, gliding distances up to 400m (1312ft), reaching a height of 6m (20ft) above the waves, at speeds exceeding 70km/h (44mph), but breaking cover can expose them to seabirds.

ALL PHOTOGRAPHS:
Gentle giant
The giant grouper (*Epinephelus lanceolatus*) is a massive fish, weighing up to 400kg (882lb), with a wide distribution through the Indo-Pacific. They are solitary, curious fish and often venture close to divers on the reef. Crustaceans are a major item in their diet, being swallowed whole by large individuals.

Flying through the ocean
Manta rays (*Mobula* species) are oceanic giants with an effective 'wingspan' of over 7m (23ft). The cephalic fins in front of the head are kept furled when the fish is swimming, and held flat as here during feeding. These rays often eat tiny planktonic creatures as well as fish.

Well-adapted
Manta rays have the largest brains of any fish, and also have a heat exchange mechanism here which may help them to stay warm when diving into the cold depths of the ocean. Their gestation period lasts just over a year, with usually one pup then being born.

ALL PHOTOGRAPHS:
A long-distance wanderer
The great white shark is probably the most feared oceanic fish, hunting mammals such as dolphins and seals, plus seabirds and fish. Its only predator is the orca. The largest recorded individual measured 5.83m (19.12ft), and weighed 2000kg (4409lb).

OVERLEAF:
Patient hunters
The hairy frogfish (*Antennarius striatus*) lives in the Indo-Pacific region and the eastern Atlantic. It has a massive mouth, allowing it to swallow prey as large as itself. It is an ambush hunter with its spinules, resembling hair, helping to disguise its presence.

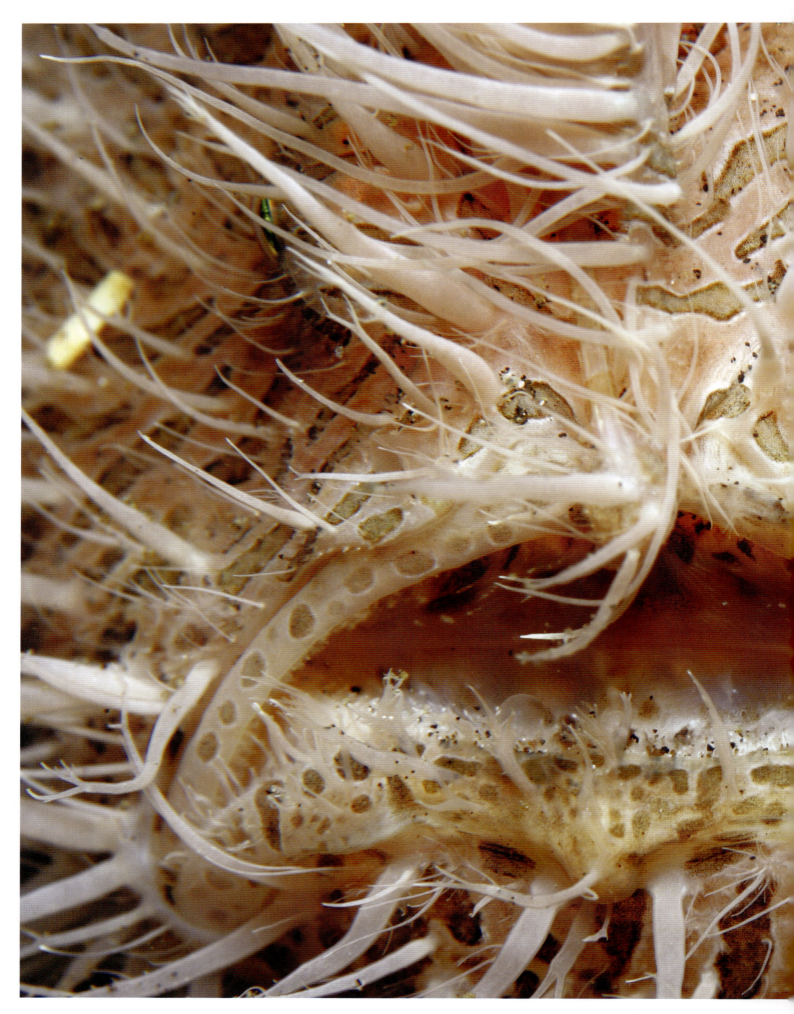

OPPOSITE:
Lightning strike
Although not able to swim fast, frogfish can dart out very quickly to capture prey that comes within reach, retaining the element of surprise. They suck the creature into their mouth and swallow it whole. Emitting a special attractant scent helps them to attract prey, notably at night.

ABOVE:
Attracted to shipwrecks
There are five living species of Platax batfish, of which t his is the orbicular batfish (*P. orbicularis*). These reef fish are found throughout the entire Indo-Pacific Ocean, and can occur in the Atlantic too. Their flattened and highly manoeuvrable bodies may measure up to 70cm (28in) in width.

ABOVE:
Effective protection
The longhorn cowfish (*Lactoria cornuta*) is an Indo-Pacific reef fish, also called the horned boxfish. While many fish are streamlined to help them swim faster, these cowfish have a rectangular body shape, but they are protected by tough body armour and can produce a potent poison.

RIGHT:
Amazing colouration
Part of the spectacular colouration of the mandarinfish (*Synchiropus splendidus*) is almost totally unique among all vertebrates. Only this species and its close relative the psychedelic mandarinfish (*S. picturatus*) have blue colouration as the result of a colour pigment in their cells. In all other known cases, the blue colour comes from thin-film interference from piles of flat, thin and reflecting purine crystals.

A defensive mechanism
The smooth appearance of mandarinfish reflects the fact that their bodies are not covered in scales, but have a layer of unpleasant-tasting slime here instead. This may protect them from potential predators, who learn to associate this unpleasant taste with their bright colouration.

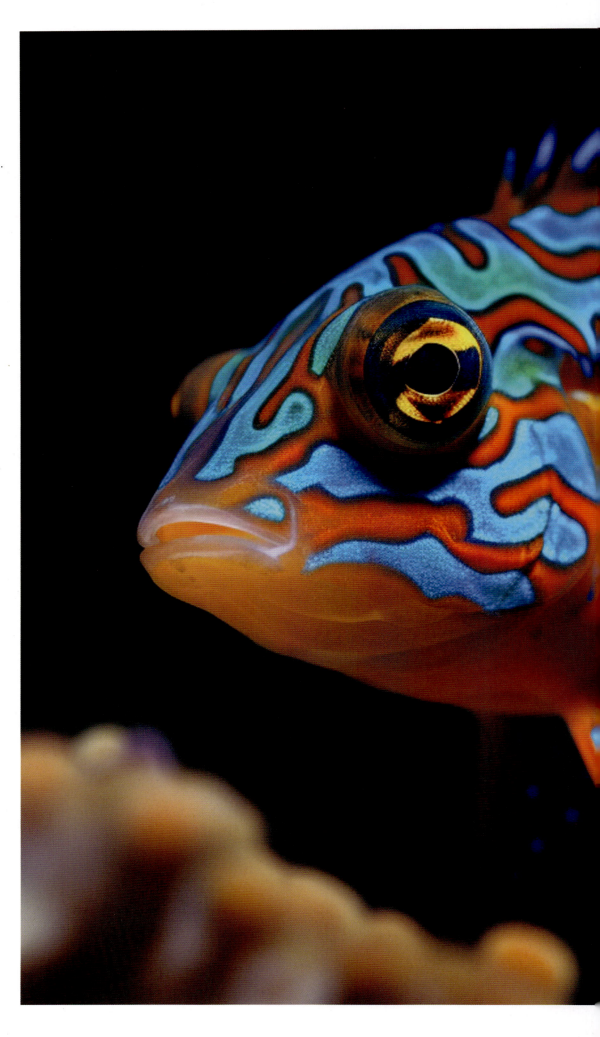

Growing up in relative safety
A view of a mangrove tree both above and below the water surface, with its roots clearly visible. Providing plenty of food and, more importantly, lots of hiding places, mangroves like this one in the Caribbean are important nurseries for young marine fish early in life.

New territory
The native habitat of lionfish (*Pterois* species) lies in the Indian Ocean and the western Pacific, although recently, they have become a significant invasive species in the Western Atlantic, including the Caribbean. The ancestry of these lionfish may trace back to aquarium stock displaced by hurricanes.

A deadly deterrent
Lionfish, like the red lionfish (*P. volitans*) seen here, are well-equipped to defend themselves, thanks to their venomous fin rays. There have even been accounts of human fatalities from their venom. Most fish therefore avoid them, although barracudas and big groupers are not deterred.

ALL PHOTOGRAPHS:
Formidable defences
Porcupinefish form a group of 18 different reef fish, well protected by spines. They can also inflate their bodies with air, have sharp teeth in their mouths, are capable of inflicting a painful bite and may contain deadly tetrodotoxin, which is far more dangerous than cyanide.

RIGHT AND OVERLEAF LEFT:
Merging into the background
An ornate ghost pipefish (*Solenostomus paradoxus*) blends into a crinoid on an Indonesian reef. These strange fish live in pairs or individually, not being seen in groups.

OVERLEAF RIGHT:
Height matters
The sunfish (*Mola mola*) is one of the largest bony fish alive today. It has a very tall body, and can measure up to 2.5m (8.2ft) vertically.

156

BOTH ABOVE PHOTOGRAPHS:
Walking not swimming
The spotted handfish (*Brachionichthys hirsutus*) is a remarkable and critically endangered Australian fish. Its highly specialized pectoral fins have evolved into hand-like structures, enabling the spotted sandfish to walk over the seabed. They are quite small fish, growing up to about 12cm (4.7in).

RIGHT:
Ready to strike
A sailfish (*Istiophorus* species) pursues a tight shoal of sardines, which are trying to outmanoeuvre their pursuer. It will use its long bill to break up the group, hoping to stun some in the process, which will make them easier to catch.

Hiding away to hunt
Although the spiny devilfish (*Inimicus didactylus*) hides away through the day, often digging itself into the sea floor, and remains disguised here, it is actually a very effective nocturnal hunter of other fish. The species possesses a potent venom, stored along the base of its dorsal fin.

ABOVE TOP:
Repurposed fins
A spiny devilfish clearly visible, out in the open. Nevertheless, well-protected by their potent venom, even predatory species do not need to engage with these fish, which grow up to 25cm (10in) long. They use their pectoral fins like legs when walking over the seabed.

ABOVE BOTTOM AND RIGHT:
The most dangerous fish
The stonefish (*Synanceia verrucosa*) is another sedentary, bottom-dwelling fish which is protected by having venomous spines. Often occurring in shallow areas of the Red Sea and the Indo-Pacific ocean, it poses a deadly threat to people who may not spot it hidden on the sea floor.

Hard to see
The varying colours and textures of the stonefish's skin help to conceal its presence, while its upward-pointing mouth helps it to seize and swallow prey such as small fish, striking in just 0.015 seconds. The fish's venom is packed into 13 dorsal spines running along its back.

ABOVE:
Easy munching
A stellate pufferfish (*Arothron stellatus*) seen here in the Red Sea. It uses the powerful teeth visible in its jaws to crush the shells of molluscs and coral polyps.

LEFT:
Different hunting strategies
The yellow trumpetfish (*Aulostomus chinensis*) may extend its mouth forwards to grab small fish or crustaceans, or it might actively stalk them, using whatever cover is available.

OPPOSITE ALL PHOTOGRAPHS:
Hiding in plain sight
The prickly leatherjack or tassled filefish (*Chaetodermis pencilligerus*) is a solitary species that relies largely on its shaggy, weed-like skin growths to disguise its appearance.

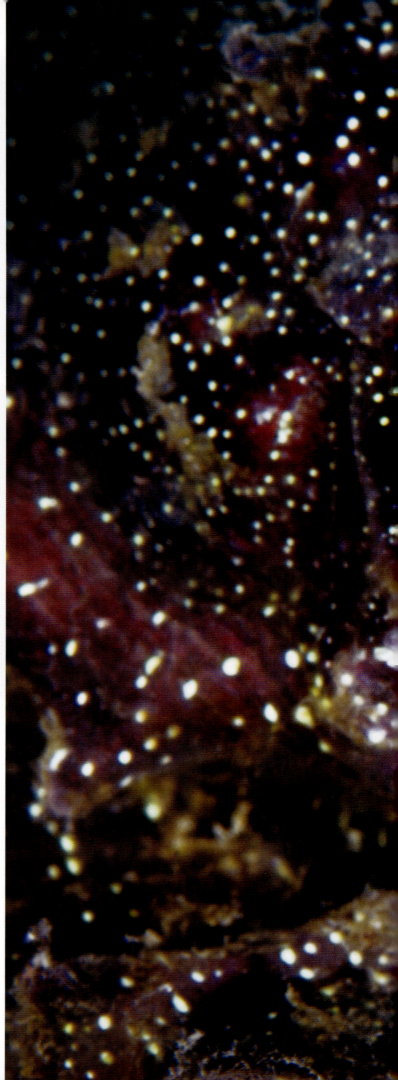

ALL PHOTOGRAPHS:
Equine appearance
Seahorses (*Hippocampus* species) can be well disguised and use their tails to anchor themselves, feeding on tiny marine creatures. The reproductive roles are reversed, with the male seahorse taking charge of the eggs and hatching them in a special brood pouch on the underside of his body.

Blending in
The weedy seadragon (*Phyllopteryx taeniolatus*) is endemic to Australia, inhabiting areas where seaweed is common, maximising the benefit of its disguise. It drifts slowly in the water, mimicking a piece of weed, and feeds on tiny crustaceans and other zooplankton. These seadragons can grow to 45cm (18in) long.

LEFT AND BELOW:
The giant of the ocean
The whale shark (*Rhincodon typus*) is the biggest living fish in the oceans today. The largest individual currently known grew to a length of 18.8m (61.7ft), and these giants can live for up to 130 years.

OPPOSITE:
An ancient survivor
The coelacanth (*Latimeria chalumnae*) was thought to have been extinct for 66 million years, up until 1952, when a living example was hauled up in a fishing net near the Comoro Islands, off Africa's east coast.

Feeding habits
In spite of its huge size, the whale shark is a filter-feeder, with a cavernous mouth which enables it effectively to trawl in search of the small planktonic organisms and small fish that make up its diet. These sharks present no danger to people.

BOTH PHOTOGRAPHS:
An unusual association
Occurring in the western Pacific, the yellow prawn goby (*Cryptocentrus cinctus*) can vary markedly in terms of its colouration, from bright yellow via brown to grey individuals. These small fish, growing to about 10cm (4in) long, live in burrows in the company of pistol shrimps (*Alpheidae*).

Dedicated parenting
A yellow goby (*Lubricogobius exiguus*) cares for its eggs on a reef off the island of Bali. These small fish, growing to a maximum length of 3cm (1.2in), are devoted parents. They are usually encountered in pairs, but are shy by nature and often hide away.

ALL PHOTOGRAPHS:
Blowing up
Pufferfish can inflate their bodies, making it much harder for a would-be predator to gain a hold, especially if they have sharp spines covering the body. Most pufferfish are marine, but a few of the 193 known species can be seen in fresh or brackish waters.

Invertebrates

Invertebrates, or 'animals without backbones', represent a large group of creatures that live in both freshwater and marine environments, although it is in the latter surroundings that they are most conspicuous and diverse. They form a vital part of the web of life in the differing aquatic environments where they occur. Zooplankton, which is a soup of tiny creatures, nourishes young marine fish and feeds much larger individuals too. Aquatic invertebrates can be divided into two broad groups, based on their lifestyles. There are the sessile invertebrates that anchor themselves to a particular spot and are then unable to more. This group includes corals and also the giant clam, with these invertebrates providing the vital basic structure of tropical reefs. The other category, sometimes described as motile invertebrates, can provide a vital source of food for other animals, not just fish, but also creatures such as marine turtles, some birds and mammals like seals. They include crustaceans, like shrimps, prawns, crabs and lobsters. The cephalopod grouping, which is confined to the marine environment, features cuttlefish, octopuses and squid, some of which can swim at considerable speed to escape danger, and represent the most intelligent members of the invertebrate group. It is incorrect to assume that all invertebrates are small: since its discovery in 2004, it is now realised that the deepwater giant squid (*Architeuthis dux*) can grow to an overall length of 13m (43ft).

OPPOSITE:
Hard to recognize
This strange creature is a sea slug known as the blue dragon (*Pteraeolidia ianthina*), which is common in the Indo-Pacific region.

LEFT:
No appetite?
The long cephalic tentacles at the head of the blue dragon sea slug have distinctive purple bands. It eats small hydroids, which contain photosynthetic organisms. These survive in the slug's digestive tract and produce sugars, so the sea slug does not need to eat to sustain itself.

ABOVE TOP AND BOTTOM:
Healthy development
Soft coral known as giant anthelia (*Anthelia glauca*), growing in the Red Sea, with a close-up view of the coral polyps which catch the small creatures that will be utilised as a source of food. There are also photosynthetic microbes in the coral tissue too.

OPPOSITE:
Breeding behaviour
Colonies of giant anthelia coral are either male or female, and spawning will take place in summer, being triggered by a full moon. Eggs and sperm are released into the sea, with the reproductive season lasting between four and five months through this period.

ABOVE:
Unexpectedly dangerous
Blue-ringed octopuses (*Hapalochlaena* species) may look attractive, but they rank among the world's most deadly marine animals. They contain the poison tetrodotoxin, which is also found in some fish, and has fatal effects on the nervous system. These grow to between 13 and 20cm (5 and 8in).

A warning
The colouration of the rings of this greater blue-ringed octopus (*H. lunulata*) are a particular intense shade of blue, suggesting that it may be stressed. The octopus will flash them for about a third of a second to emphasize the danger of an attack.

The biggest form
The giant Australian cuttlefish (*Sepia apama*) is the world's largest cuttlefish, potentially attaining a total length of more than 100cm (39in), including its extended tentacles, and individuals can weigh more than 10.5kg (23lb). In terms of their physiology, cuttlefish have three hearts in their bodies, and blue blood.

Anatomy and lifestyle
The rear of the cuttlefish's body is called the mantle, while attaching to the head are its eight arms and a pair of feeding tentacles. Cuttlefish are quite short-lived, with a life expectancy of 1–2 years. This group is spawning, after which they will then die.

Recognizing danger
A spotted jellyfish (*Mastigias papua*) swims over a brain coral. The coral represents no threat to the jellyfish, but it is at risk from sea anemones of the genus *Entacmaea* which use their tentacles to catch these relatively small jellyfish, which may only measure 10cm (4in) in diameter.

A remarkable lifespan
Brain coral seen in close-up. This type of coral is so-called because its rounded shape and folds are somewhat reminiscent of the human brain. It has a unusually long lifespan, with some brain coral colonies known to be more than 900 years old.

ABOVE:
Deep-water dwellers
Giant isopods (*Bathynomus* species) occur in the depths of the ocean, where the water is cold. They are related to terrestrial woodlice (pill bugs) and, similarly, they can curl into a ball.

RIGHT AND OPPOSITE:
Differing environments
Colourful invertebrates can be encountered in freshwater localities, too, like this cardinal shrimp (*Caridina dennerli*), from Sulawesi's Lake Matano. The larger female carries her eggs under her abdomen.

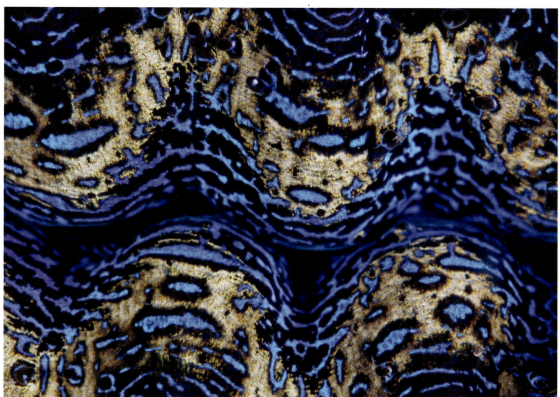

ALL PHOTOGRAPHS:
Clamming up
The beautiful internal mantle of a giant clam (*Tridacna maxima*), which may grow to 1.4m (4.5ft) long. As bivalves, these clams can slowly close their shell up if disturbed. They obtain nutrients both by filter-feeding and by the presence of photosynthetic algae in their bodies.

BOTH PHOTOGRAPHS:
Burrowing shrimps
Peacock mantis shrimps (*Odontodactylus scyllarus*) are found throughout the Indo-Pacific region and can grow to about 18cm (7in) long. They burrow into the bases of coral reefs, and catch a variety of invertebrate prey, possessing immense power to punch their way in through the shell.

Remarkable power
A head study of a peacock mantis shrimp. The power of a punch from one of these shrimps is remarkable. Its hammer-claw is reputedly strong enough to break aquarium glass. It has the quickest punch of any known animal, with its speed estimated to be 80km/h (50mph).

Colourful crabs
Sally Lightfoot crabs (*Grapsus grapsus*) are found on the coasts of South America, living among the rocks. Adult crabs can vary in colour, whereas the young are dark brown or black at first. They naturally eat algae, but will also sample a wide range of other food.

LEFT:
Maintaining a balance
Sally Lightfoot crabs being soaked by waves on a rock on the coast of North Seymour Island, within the Galápagos Archipelago. These crustaceans are surprisingly agile, gripping on to the rough volcanic stone with their legs, and they can climb up vertical surfaces without difficulty.

OVERLEAF:
Not what they seem
In spite of their name, porcelain crabs, which are members of the *Porcellanidae* family, are not crabs at all, but are related to squat lobsters. They are fragile – like porcelain! – and can frequently lose their claws, although these will grow back gradually when they moult.

LEFT AND BOTH OVERLEAF:
Crawling around
There are more than 2000 species of spiny brittle stars known today, just over half of which are found in deep water, below 200m (656ft). They crawl along over the sea bed, using their distinctively long, slender and relatively fragile arms, which set them apart from true starfish.

215

ALL PHOTOGRAPHS:
Wake-up call
Striped pyjama squid (*Sepioloidea lineolata*) typically do not grow larger than 7.5cm (3in) but they are both toxic, thanks to glands under the body that produce a unpleasant slime, and venomous. As a third method of defence, they can release a cloud of ink into the water.

ALL PHOTOGRAPHS:
Marine snails
Tiger cowrie shells (*Cypraea tigris*) can reach about 15cm (6in) long. They are home to a large form of sea snail. Adult tiger cowries will feed on coral colonies and other invertebrates that they can catch while roaming over the reef. The young are vegetarian, feeding on marine algae.

Stunning patterning
The beautiful shell pattering of tiger cowrie shells is quite individual, and in the case of juveniles, it consists of darker jagged lines on a light background rather than spots. The underside of the shell is white, with tooth-like serrations over the entry point to the shell.

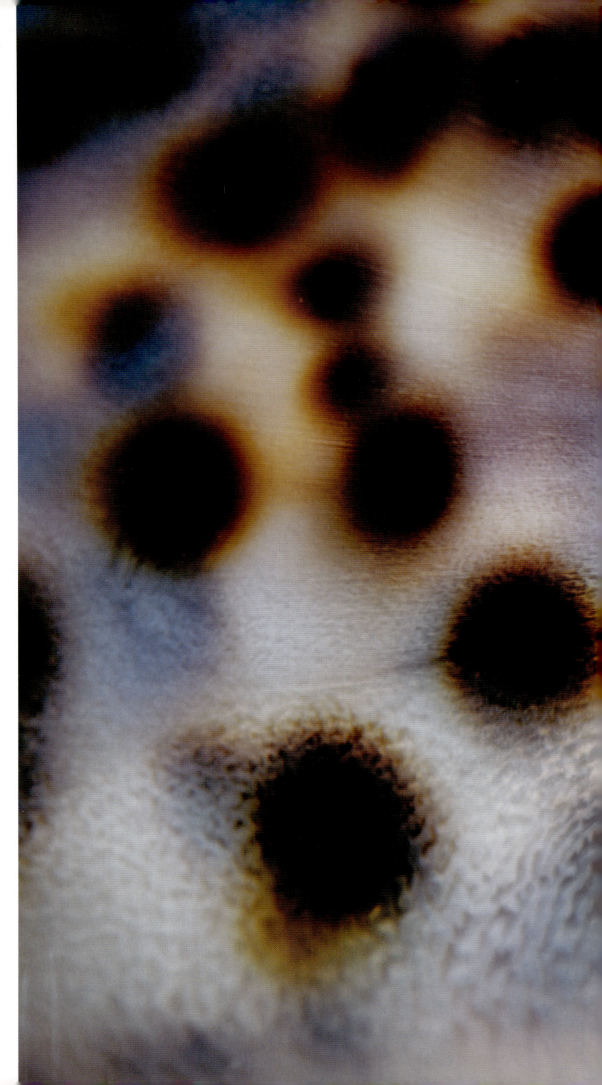

Picture Credits

Alamy: 12/13 (Robert S. Michelson), 14/15 (A & J Visage), 23 (Amar and Isabelle Guillen), 30 (Panther Media GmbH), 31 (Maximilian Weinzierl), 58/59 (blickwinkel), 61 (Juniors Bildarchiv GmbH), 64 (WaterFrame), 66/67 (imageBroker.com GmbH), 76 bottom (Wildlife GmbH), 78 top (imageBroker.com GmbH), 86 (SBS Eclectic Images), 87 top & 90/91 (blickwinkel), 96 (Helmut Corneli), 98/99 (Doug Perrine), 105 top (Marek Stefunko), 106 top (WaterFrame), 106 bottom (Animal Stock), 124 top (Francesco Ricciardi), 124 bottom (Oksana Maksymova), 125 (Erika Antoniazzo), 128 (Ross-Tom Stack Assoc), 129 top (WaterFrame), 129 bottom (Nature Picture Library), 136/137 (Luiz Puntel), 154 top (imageBroker.com GmbH), 154 bottom (Life on white), 155 (Marli Wakeling), 160 top (Nature Picture Library), 161 (Connect Images), 162/163 (imageBroker.com GmbH), 164 bottom (imageBroker.com GmbH), 165 (Ethan Daniels), 166/167 (Charles Stirling Diving), 175 top (imageBroker.com GmbH), 175 bottom (Nature Picture Library), 176/177 (Reinhard Dirscherl), 180/181 (Blue Planet Archive), 182 (RooM the Agency), 183 top (Matthew Oldfield Underwater Photography), 183 bottom (imageBroker.com GmbH), 187 top (imageBroker.com GmbH), 188 (Marli Wakeling), 200 top, 201 & 210/211 (BiosPhoto), 212/213 (Underwater), 214/215 (WaterFrame), 216 (Marli Wakeling), 218 (Andrew Trevor-Jones), 219 top (Andrew Trevor-Jones), 219 bottom (Nature Picture Library)

Dreamstime: 5 (Photographerlondon), 8 (Joan Carles Juarez), 16/17 (MisterTigga), 18 (Galinasavina), 19 top (Kurniawandeny20), 19 bottom (Astatotilapia burtoni), 22 top (Verastuchelova), 22 bottom (Hamsterman), 24/25 (Valeronio), 26/27 (Copora), 28/29 (Cookelma), 32 top (Dzombie), 32 bottom (Miropa), 33 (Kinnon), 35 bottom (Mirkorosenau), 40/41 (Slowmotiongli), 42/43 (Mirkorosenau), 44 bottom (Diegograndi), 45 (Miropa), 47 bottom (Diegograndi), 48 left (Wrangel), 48/49 (Copora), 54 (Jcjuarez9), 55 top (Mirkorosenau), 55 middle (Joan Carles Juarez), 55 bottom (Mirkorosenau), 56/57 (Slowmotiongli), 60 top (Allexxandar), 60 bottom (Allexxandar), 65 top (Slowmotiongli), 68-69 (Marcinadrian), 78 bottom (Wrangel), 79 (Vojcekolevski), 83 (Marcinadrian), 92/93 (Mikhailg), 104 bottom (Gonepaddling), 107 (Ethanadaniels), 110/111 (Mirecca), 120 (Daexto), 121 top (Cbpix), 126 bottom (Vlad1949), 127 (Grace5648), 138 bottom (Ramoncarretero), 140/141 (Johnandersonphoto), 142 (Kelpfish), 146/147 (ArtesiaID), 150/151 (Crisod), 156/157 (Ethanadaniels), 160 bottom (Rixie), 169 top (Mirecca), 169 bottom (Djmattaar), 172/173 (Kristinapchel), 174 (Slowmotiongli), 178 (Vojcekolevski), 190/191 (Pics516), 196/197 (Seadam), 198/199 (Pniesen), 203 bottom (Ethanadaniels), 205 (Kjorgen), 217 (Panther0550)

Shutterstock: 6 (Thierry Eidenweil), 7 (frantic00), 10/11 (smutan), 15 (Daboost), 20 top (Guillermo Guerao Serra), 20 bottom (Ttonn), 21 (ArtEvent ET), 34 top (MatMeyers), 34 bottom (nounours), 35 top (Photofenik), 36/37 (Pavaphon Supanantananont), 38/39 (Arunee Rodloy), 44 top (Opayaza12), 46 (Evgenii Predybailo), 47 top (Diego Grandi), 50/51 (Arunee Rodloy), 52/53 (Gonzalo Jara), 62/63 (Danny Ye), 65 bottom (Maris Grunskis), 70/71 (Mark Green), 72/73 (Kateryna Moroz), 74/75 (tupulointi), 76 top (Dany Kurniawan), 77 (ShaunWilkinson), 80/81 (Neryxcom), 82 top (Galliina), 82 bottom (kmn sandamali), 84/85 (Guillermo Guerao Serra), 87 bottom (Pavaphon Supanantananont), 88/89 (Toxotes Hun-Gabor Horvath), 94/95 (topimages), 100 (SergeUWPhoto), 101 (marcobrivio.photography), 102/103 (sirtravelalot), 104 top (Joe Belanger), 105 bottom (Marut Sayannikroth), 108/109 (randi ang), 112/113 (blue-sea.cz), 114/115 (Vlad61), 116/117 (Subphoto.com), 118/119 (Foodies Academy), 121 (Richard Whitcombe), 122/123 (Nantawat Chotsuwan), 126 top (Paul Atkinson), 130 (Vallehr), 131 (Daniel Huebner), 132 (Martin Prochazkacz), 133 top (Prill), 133 bottom (aqustudio), 134/135 (Hoiseung Jung), 138 top (Sergey Uryadnikov), 139 (Shane Myers Photography), 143 (Minakryn Ruslan), 144 (asawinimages), 145 (Kurit afshen), 148/149 (Damsea), 152/153 (Michael Siluk), 158 (John Back), 159 (Tomas Kotouc), 164 top (DiveSpin.Com), 168 all (Pavaphon Supanantananont), 170 top (Frolova Elena), 170 bottom (Francesco Ricciardi), 171 (Kristina Vackova), 179 (Kurit afshen), 184 (Ethan Daniels), 186 (xlchen), 187 bottom (Dray van Beeck), 189 (QJ Kang), 192/193 (Philip Garner), 194/195 (wildestanimal), 200 bottom (Shrimplake), 202 (kai egan), 203 top (Ethan Daniels), 204 (Gerald Robert Fischer), 206/207 (Mike Workman), 208/209 (nwdph), 220 top (Niklas Schubert Rocha), 220 bottom (RobJ808), 221 (Vlad61), 222/223 (Ethan Daniels)